Carlos M. Carvalho de Jesus

From excess food to food shortages in the society of abundance

Carlos M. Carvalho de Jesus

From excess food to food shortages in the society of abundance

The fight against food waste

ScienciaScripts

Imprint
Any brand names and product names mentioned in this book are subject to trademark, brand or patent protection and are trademarks or registered trademarks of their respective holders. The use of brand names, product names, common names, trade names, product descriptions etc. even without a particular marking in this work is in no way to be construed to mean that such names may be regarded as unrestricted in respect of trademark and brand protection legislation and could thus be used by anyone.

Cover image: www.ingimage.com

This book is a translation from the original published under ISBN 978-613-9-72529-8.

Publisher:
Sciencia Scripts
is a trademark of
Dodo Books Indian Ocean Ltd. and OmniScriptum S.R.L publishing group

120 High Road, East Finchley, London, N2 9ED, United Kingdom
Str. Armeneasca 28/1, office 1, Chisinau MD-2012, Republic of Moldova, Europe
Printed at: see last page
ISBN: 978-620-7-92331-1

To Prof Dr Iva Pires, for her willingness to accompany my work, for her criticism and encouragement.
To the tireless Carla and dear sons Diogo and Gonçalo, for their constant presence and support.
To my brother Adriano for all his support.
To my colleagues from the Human Ecology Master's programme, especially Pedro Brum, for the moments and the memories that remain.
To the FCSH and all its teaching staff, for everything they have taught me over the last few years. Thank you!
To Hunter Halder, Pedro Almeida and Célia Casinha, Isabel Soares and Maria and all those who collaborated with me at the Re-Food centre and at Fruta Feia.
To my father, Ângelo de Jesus, for making sure that this would be a real source of joy and pride.

Thank you all very much!

Summary
Sharing the guidelines and concerns of the United Nations Organisation, expressed in 2015 in the Sustainable Development Goals and whose main goals are to "End hunger, achieve food security and improved nutrition and promote sustainable agriculture" (SDG2) by 2030 and the need to "Ensure sustainable production and consumption patterns" (SDG12), we propose to reflect on the problem of combating food waste and on social actions aimed at mitigating it. It is interesting to understand the socially constructed relationship between overconsumption and food waste on the one hand, and food shortages on the other, as this seems to generate social inequalities and injustices.
With a view to the Circular Economy and using the various levels of the food waste hierarchy, we will learn about practices and initiatives created at each level, always favouring prevention/reduction and food donation, as they are those that can be adopted before food becomes food waste, and are therefore to be favoured from a social, economic and environmental perspective.
In Portugal, we will analyse two institutions that, with different modi operandi, have set out to combat food waste. The **"Fruta Feia"** consumer cooperative**, an entrepreneurial project that aims to combat market inefficiency** by putting back on the market products that have been rejected by the commercial circuit for aesthetic or calibre reasons, and the **"Re-Food 4 Good"** Association**, a** micro-local **organisation** made up of volunteers that works daily to reduce food waste and hunger in each neighbourhood, while strengthening community ties.

Key words: food waste, food redistribution, circular economy, sustainable development

Index

CHAPTER 1

Introduction

Today we live in a network society, in a global, uncertain and constantly changing world, an era characterised by growing interdependence, interconnectivity and great complexity, which have made social, environmental and economic systems extremely vulnerable on a global scale, generating systemic risks with unpredictable consequences.

In reality, there have been many profound transformations in communities and societies, changes that have been the subject of the most diverse approaches since the 19th and 20th centuries.

From the classic authors of sociology (Simmel, Durkheim) when they analysed the different ways of life - the traditional, with communal social relations, and the modern with individualisation, its main characteristic - to contemporary authors (Bauman, Giddens, Castells), they have all looked at the differences between community and society and the basic characteristics of societies that promote or inhibit their development.

Georg Simmel, one of the forerunners of the Chicago school in the first quarter of the 20th century, defined the city and the metropolis as the "cradle" of individuality, where the presence of impersonality and highly differentiated subjectivity is great. In the metropolis, indifference emerges, "the blasé attitude", i.e. an inability to react to stimuli and new sensations.

In Émile Durkheim's work on the division of social labour we find a formulation of the idea of community. His main concern, however, was with the radicalisation of the division of labour from the 19th century onwards and its general consequences for societies where the greater division of labour came to mean greater specialisation. If, on the one hand, this led to greater individualisation of individuals, through their differentiation from the majority, on the other hand, it led to greater integration, through interdependence between people who have skills that other individuals need.

However, Durkheim warned of the risk of anomie, i.e. the lack of integration and norms that regulate social life. Anomie is the result of rapid social change since the Industrial Revolution (and the French Revolution) and seems to be related to the crisis of traditional values and the inability to integrate socially in the modern world. However, for him, the normal state of contemporary society, as opposed to traditional society where mechanical solidarity prevailed, is that of organic solidarity, i.e. a new form of integration. However, anomie would not be permanent and society would be able to reorganise itself and eliminate what is harmful to it.

For Anthony Giddens, a contemporary author, this anomie is the result of a discontinuity between traditional social orders and modern social institutions that has jeopardised the social pact itself.

Zygmunt Bauman has reflected on the changes that modern society has gone through, from individualism to work, family and community relations, where time and space are no longer concrete and absolute but "liquid" and relative, and where he emphasises "the fragility of human bonds" in this modern society. "Liquid modernity is a civilisation of excess, superfluity, waste and its removal." (Bauman, 2005:120).

He analyses emancipation, individuality, work and community. For him, identity is the contemporary substitute for community. Faced with the impossibility of finding a community, a new form of understanding and security, the concept of identity now gains importance. The individual achieves freedom of action and thought, in the continuous search for fulfilment and self-affirmation, breaking ties with the environment.

Today in the West we have a modern, open society, a type of society where the increase in production has led to the multiplication of consumer goods and, consequently, the creation of new needs and desires, a phenomenon that invades the spheres of social, economic and cultural life and which is orientating the whole symbolic dynamic of society.

Authors such as Bauman, Baudrillard and Lipovetsky state that in a consumer society, "having" becomes more important than "being". Those who don't consume are left out of the social sphere, which has given rise to a new form of social exclusion and where "human solidarity is the first casualty caused by the triumph of the consumer market" (Bauman, 2004:96).

Bauman links quality of life to consumption. In this sense, he states that this is a way of perceiving quality of life and emphasises that it is impossible to create a world in which everyone obtains the desired quality of life. The author argues that this is an unrealistic goal, because if quality of life is defined by consumption, it would take three Earths just like the one we live on to guarantee quality of life for everyone. This is because, at the moment, "the better the quality of life, the greater the ecological footprint left by a city on the planet we share" (Bauman, 2005), meaning that the ecological issue, in the consumerist logic, has no room for discussion

For Baudrillard (2008), today's society is organised on the basis of consumption, and no longer on the basis of production (...) The consumer's desire for goods is not directed at the objects themselves, but rather a desire for inclusion within the consumer system.

Lipovetsky (1988), when analysing contemporary society and contemporary individualism, highlights the weakening of society and customs in the era of (hyper)mass consumption where "we are all destined to consume".

For his part, Castells (2007), when analysing contemporary society and its social morphology, identifies a new social structure and a new social organisation "the network society" marked by the presence of a system of interconnected networks that has transformed social relations, value systems and the very mode of development. A process whose core transformations lie in globalisation and the technological revolution, with the use and application of information and knowledge and which, due to the enormous social changes it has caused, has simultaneously proved to be a mechanism for social inclusion and exclusion.

On the other hand, and in response to the increase in population, initially with the movement of people from the countryside to the cities, and later with globalisation and space-time compression (Harvey, 1996), the entire social space has been redefined.

New and multiple social relationships have emerged which, in the view of Giddens (1994), have led to a rise in the phenomenon of "reflexivity" in society, a phenomenon that has been mirrored in the daily lives of individuals and in the most varied dimensions (social, cultural, labour, environmental, food, etc). Social reflexivity, experienced and felt in a context of globalisation, economic growth and a (hyper) consumer society, could only lead to major societal changes.

Social transformations that have cut across all areas and sectors of society, as is the case with food, a complex, multifaceted and multidimensional phenomenon, a cultural instance that preserves customs, which is a pillar of people's cultural identity and essential for the survival and well-being of the individual.

In the context of (food) Modernity and increasing Globalisation, it is possible to identify changes, behaviours and impacts that food (the agri-food system) and the act of eating have undergone as a result of the social, economic and cultural transformations that have taken place in contemporary society, particularly the act of wasting food.

In fact, it can now be seen that the growth in food production associated with a better quality of life and well-being is causing a huge amount of food waste.

An unsustainable and immoral situation, since the needs of a growing world population need to be met with essential resources that are becoming increasingly scarce.

According to UN forecasts, the world population will increase from the current 7.3 billion people to 9.7 billion by 2050, mostly living in urban areas, with the FAO estimating that if the current rate of consumption continues, the world will need 70 per cent more food[1] , which will mean consuming 50 per cent more energy and 40 per cent more water, so agricultural and food systems must do more with less (IFPRI 2015).

Sharing the guidelines and concerns of the United Nations Organisation, expressed in 2015 in the Sustainable Development Goals and whose main targets are, by 2030, "End hunger, achieve food security and improved nutrition and promote sustainable agriculture" (SDG2) and the need to "Ensure sustainable production and consumption patterns" (SDG12), we propose to reflect on the problem of food waste and social actions aimed at mitigating it.

[1] The arrival of emerging economies (countries like China or India whose economies are now in full economic development) in the capitalist mode will have a huge impact on the need for food and on food security itself (Pollan, 2009). According to FAO forecasts, world meat consumption is expected to rise from 32 to 52kg per capita per year (FAO, 2011).

CHAPTER 2

Food Waste. From lack to waste

> "Every day nature produces enough for our needs. If everyone took what they needed, there would be no poverty in the world and no one would starve." Mahatma Gandhi

[a]Food waste, a phenomenon that emerged after the end of World War II, coincided with the start of the post-war cycle of economic growth and consumerism. A situation that contrasted with the previous era, when food rationing and hunger accompanied the reality of war, which is why food was highly valued (Figure I) and religiously respected. "In the world the greatest wealth, For those who think well, Is to put bread on the table, Of those who have none" (Bread of the poor, verse dedicated to St Anthony of Lisbon)

Figure 1 - World War II posters [a]

Food is the basis of life and human health and well-being depend on it. Every day, around the world, large quantities of food that could have been eaten are wasted, which is why food waste, a global (and local) problem, has become a political and social priority in recent years.

Although food production is currently sufficient to feed the world's population, the return of food security concerns[2] suggests that reducing food waste should become a serious objective for sustainable development.

In 2017, around 821 million people, 11% of the world's population, suffered from hunger, a figure higher than in 2016 (FAO, 2018). At the same time, due to malnutrition, 641 million people (13% of all adults on the planet) suffer from obesity, a phenomenon already classified by the World Health Organisation as the epidemic of the 21st century, which led Raj Patel (2008) to state that "Global hunger and overweight are symptoms of the same problem".

It should be emphasised that access to food and the quality of food are key issues in human development, a concept that points to social progress.

> **"Starving to death is characteristic of some people who do not have enough food to eat. It is not characteristic of not having enough food to eat" (Sen, 1999:II).**

[2] We can define food security based on four pillars: availability, access, adequate utilisation of food and stability. Conversely, when access to food is reduced or non-existent (hunger) and inadequate from a nutritional point of view, we have food insecurity.

With the growth of consumer society, the modern food industry has changed the way we consume. In its eagerness to produce more for more people, it has turned food into a commodity. It has transformed our subjectivity and our food identity, our lifestyles and the ways we relate to others and to things (from dressing to eating). In short, it has transformed the entire social space.

New food logics have replaced traditional food supply systems, causing individuals to develop new habits and behaviours.

With the deterritorialisation of food (and services) on a global scale, since the end of the 20th century we have seen a certain "uprooting of food" (Ortiz, 1994) as a result of the diversification of products and the shift from traditional to industrial cuisine. New food products have been created by the food industry characterised by an excess of artificial products, to the detriment of the consumption of regional products with a strong cultural tradition. We have moved away from the Mediterranean diet.

We've gone from a diet based on cereals, tubers, with low levels of lipids and few animal products, to consuming processed foods, which are cheaper but full of fats and sugars, high in calories, from home-made food to outsourced food, made by others, in which family relationships, cultural identity and ethical diversity, which were closely related to the act of preparing and consuming food, are now the responsibility of large corporations (Roberts, 2008).

> "Food (globalised and industrial agriculture) is the biggest health problem in the world, and also the biggest problem for the health of the planet." (Vandana Shiva, 2018)[3]

Nowadays, with the lengthening of supply chains, putting more and more distance between the producer and the consumer, which means more intermediary agents in the chain and greater transport needs, food (merchandise) produced out of season and in traditional locations is accessible to everyone at any time, completely breaking with a food tradition that was orientated around proximity[4] .

That's why issues such as the ecological footprint and the sustainability of our production model - from processing, transport, distribution to food consumption - will always be part of the food debate.

As a result of the growth of cities and new ways of life, the act of eating has expanded into public spaces, after being confined to the home for a long time. Factors such as the extension

---

[3] http://www.ihu.unisinos.br/78-noticias/578380-a-comida-e-o-maior-problema-de-saude-que-ha-no-world-interview-with-vandana-shiva

[4] "Food miles" is the term used to calculate the distance (and environmental impacts) travelled by food throughout the process, from production to reaching our plate.

of the domain of work and its separation from family life, as well as the entry of women into the labour market, have contributed to new ways of life which, combined with the growth and development of food technology, have opened up space for the food industry, restaurants, department stores and multinationals.

According to Roberts, we have lost control of our diet, which is why we need to reflect and act on the current food system and our eating behaviour.

New social dynamics associated with the abundance and diversity of food created[5] , have led to the emergence of new consumption contexts and new lifestyles, but in a negative sense, whether due to the aggressiveness/competitiveness of the food industry or the excessive and somewhat careless consumption of each individual, this has made enormous food waste visible throughout the food chain.

It is a food system that has failed to respond to growing food insecurity (Gustavsson et al. 2011; Lundqvist et al.2008) but which has strongly contributed to the depletion of the planet's natural resources, which are necessary for the next generations, and to environmental degradation (deforestation), which has made the need for economic growth that points towards sustainable development increasingly urgent[6] which, as Pope Francis emphasises, considers an "integral ecology", i.e. the need to have an integrated and non-fragmented view of human beings, society and nature (Encyclical Letter Laudato Si' 2015).

> The perception of high environmental impacts and the ethical issues arising from unequal access to goods and services in the context of the consumer society have led to the questioning of the current model of consumption, especially that of the more industrialised countries, and the discussion of alternative models of sustainable consumption (Baudrillard, 2008).

We are therefore facing a problem of sustainability because, in addition to the useless expenditure of environmental and economic resources caused by this waste, from a moral point of view we cannot remain indifferent. The disparity between food waste and food shortages draws attention to the social and ethical implications of food surpluses.

In addition to the unequal distribution of food around the world, another problem we can add is society's behaviour towards food and food waste, a situation that needs to be rectified.

There are more than enough reasons to draw our attention to making more efficient use of

[5] Never in history has the population of developed societies had so much to eat and been so free from hunger as now. Today, according to Fischler (1995), in the minds of contemporary diners, the crucial question is increasingly what to eat and in what proportion.

[6] Sustainable development is the rational use of natural resources in favour of social well-being, guaranteeing the economic growth necessary to meet our demands and the needs of future generations.

food and the need to combat and reduce food waste.

CHAPTER 3

How do you define food waste?

Food is produced with the aim of being consumed. Food that does not fulfil this purpose constitutes "food waste", a situation that, according to the FAO, occurs at every stage of the food chain, from production to the final consumer.

As well as representing an extremely inefficient use of the planet's natural resources, it entails heavy economic costs (the cost related to the value of the products themselves and the costs incurred in producing, transporting and storing the wasted products, as well as treating the waste) and environmental costs (soil, water, energy, increased greenhouse gas emissions, impact on biodiversity).

However, despite the growing importance of food waste on the political and social agenda in recent years[7] , the real scale of the problem is unknown. Different studies present divergent data for each sector of the food supply chain.

The existence of different definitions for food waste, as well as the lack of a common methodology for measuring the level of this waste (edible/inedible component), has hindered progress towards understanding the real dimensions of the problem, which led the FAO to distinguish between "food losses" and "food waste".

Food loss is the unintentional reduction of food available for human consumption, which occurs mainly in developing countries (around 54 per cent) at the production, harvesting and processing stages and is a natural result of inefficiencies in production and industrial systems (lack of adequate storage infrastructure, shortage of refrigeration equipment, lack of technology and low investment in agricultural production systems) and weather conditions that leave many crops in the field.

On the other hand, food waste (around 46%) is an avoidable loss that occurs above all in more developed countries, where there is a trend towards globalisation and centralisation of the food sector, in the distribution and consumption phases, and refers to food, sometimes in good condition, that is thrown away, caused by rules and regulations of the commercial circuit (e.g. tight expiry dates or products excluded due to aesthetic requirements, etc.) and by consumer behavioural reasons (e.g. too much food bought/cooked, etc.). tight expiry dates or products

[7] This has been a concern since 1974, when the 1st World Food Conference ([a]) was held, where reducing post-harvest waste was considered a priority for development (Rome, 1974).

excluded due to aesthetic requirements, etc.) and for behavioural reasons on the part of consumers (e.g. poor food preservation, too much food bought/cooked, cultural habits, etc.).

There is no agreed definition of food waste at EU and Member State level either. For the European Court of Auditors, "food waste" refers to any product or part of a product grown, fished or processed for human consumption that could have been consumed if it had been treated or stored differently (ECA, 2016).

CHAPTER 4

How much is wasted?

According to the Food and Agriculture Organisation of the United Nations (FAO), around a third (1.3 million tonnes) of all food intended for human consumption is wasted every day between the place of production and consumption, which could feed the more than 800 million people who go hungry, approximately 30% of the world's agricultural land, with consequent environmental and economic costs[8] , so we question whether it isn't more rational and more sustainable to make an effort to reduce the high volume of food waste, rather than just looking for new land to increase food production.

In the United States, estimates of food waste in distribution point to significant figures, such as 30 per cent of fruit and 32 per cent of fresh vegetables (NRDC, 2013).

A concern that is fully recognised in Europe,

> **"Food waste is a major problem on a global scale and is a phenomenon typical** of **consumer** society **in** the **Western world" (European Commission 2014).**

In the UK, according to the TESCO chain, of the 71,000 tonnes of surplus, 38,000 were still suitable for human consumption, most of which went to animal feed and energy production.

In Denmark, the agri-food industry generates around 716,000 tonnes of avoidable food waste every year, 133,000 tonnes of which is in the processing sector, despite this being the most efficient sector in the food supply chain.

Looking at Figure II, we can see that the contribution of each region is very uneven. North America and Europe are the regions where per capita food waste is highest and sub-Saharan Africa and the South East where it is lowest (FAO, 2011). With regard to losses, it can be seen that in Africa they occur in the early stages, which substantially reduces the availability of food in these regions, one of the reasons given for the strong food insecurity there. In Europe, wastage is widely perceived in the final stages, mainly at consumer and family level.

[8] Measured in weight, FAO, Global food losses and food waste - extent, causes and prevention, Rome, 2011.

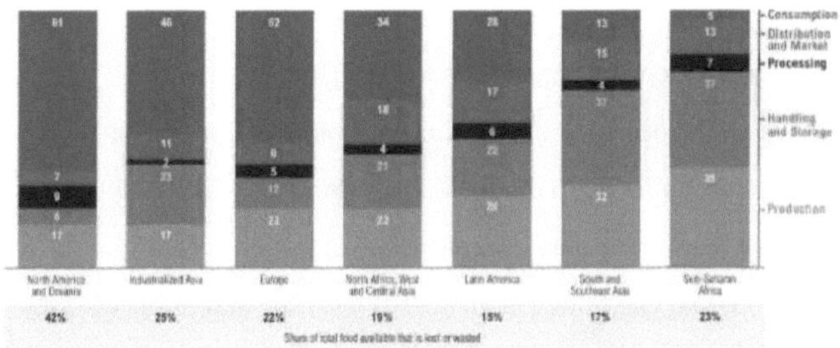

Figure II - Percentage of kcal lost and wasted. Source: WRI analysis based on FAO 2011

According to the FAO, food produced but not consumed uses a volume of water (250 cubic kilometres) and is responsible for 8% of the world's greenhouse gas emissions, as well as contributing to the loss of biodiversity.

When it comes to greenhouse gas emissions, not all foods have the same impact. As we can see in Figure III, vegetarian and pescetarian diets have significantly lower carbon dioxide emissions than diets in which pork, poultry and beef are frequently on the menu, such as the Mediterranean diet.

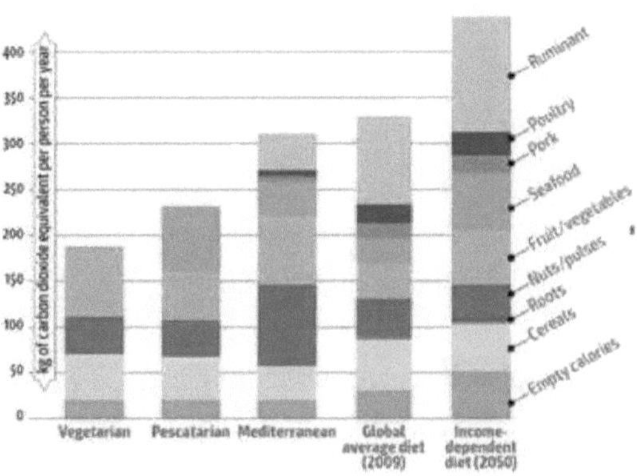

Figure III - Relationship between GHG emissions and type of diet (IFPRI, 2015)

But the impact of food on the environment will probably become more pronounced in the coming decades, as more people move into the middle class (as is the case in emerging economies) and start consuming meat-based diets (Income-dependent diet).

Today, 20% of the 263 million tonnes of meat produced on the planet is wasted, and the livestock sector is one of the biggest contributors to greenhouse gas emissions. Meat production, with its emissions of methane into the atmosphere, a gas 25 times more powerful than carbon dioxide (FAO, 2013), contributes 18% to the increase in global warming, a figure higher than that of transport (13.5%).[9]

In short, the way we currently produce and consume is not sustainable.

A study carried out in 2013, on the occasion of World Environment Day, by the World Resources Institute (WRI) and the United Nations Environment Programme (UNEP), pointed out that "One in four calories produced by the world's agricultural system is lost or wasted"

According to Lundqvist (2008) and Guyomard et al. (2012) less than half of the calories produced globally by farmers reach our tables, i.e. of the 4,600 kcal produced daily by each inhabitant of the world, only 2,000 kcal are actually consumed by people. (Guyomard et al, 2012)

In the European Union, it is estimated that around 88 million tonnes are produced each year.

tonnes of food waste (180 kg per person), a large proportion of which is food that is still fit for human consumption (EU FUSIONS, 2016)[10] , it is estimated that, if no action or preventive measures are taken, this figure will rise to approximately 126 million tonnes by 2020[11] .

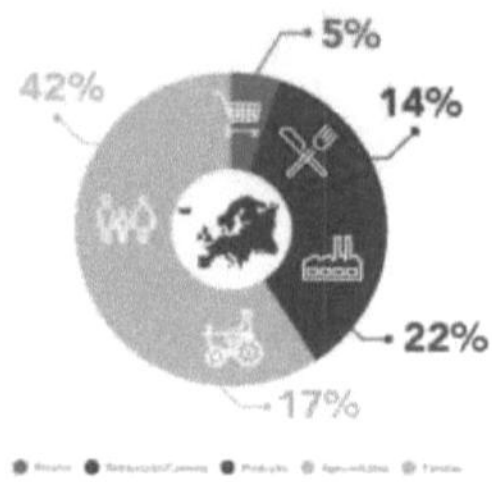

Figure IV - Breakdown of food waste by sector of activity in Europe (Eurostat, 2008)

In the European Union, it is estimated that around 88 million tonnes of food waste (180 kg per person) are produced every year, a large proportion of which is food that is still fit for human consumption (EU FUSIONS, 2016)[12] , and it is estimated that, if no action or preventative

[9] food consumption is the main contributor to global warming in Europe (31 per cent), surpassing housing (23.6 per cent) and transport (18.5 per cent) (Tukker A et al., 2006).

[10] Available at FUSIONS Reducing food waste through social innovation; http://www.eu-fusions.org/phocadownload/Publications/Estimates%20of%20European%20food%20waste%20levels.pdf

[11] European Commission, Directorate-General for the Environment, Preparatory study on food waste across EU 27, 2010.

[12] Available at FUSIONS Reducing food waste through social innovation; http://www.eu-fusions.org/phocadownload/Publications/Estimates%20of%20European%20food%20waste%20levels.pdf

measures are taken, this figure will rise to approximately 126 million tonnes by 2020[13] .
In Portugal, the study by PERDA - Projecto de Estudo e Reflexão sobre Desperdício Alimentar (Project for the Study and Reflection on Food Waste)[14] , estimated that along the entire chain, waste represents 17 per cent of annual food production, around 97 kg per inhabitant per year, of which 31 per cent comes from consumers (Baptista, et.al., 2012).

In the Western world "(- ■-) we eat a lot, we eat badly and we waste a lot of food" (Batista, et al, 2012).

In terms of economic impact, food waste represents high waste management costs and wasted money, given the considerable amount of edible food wasted every year in the EU. Such waste management costs include the maintenance of landfills (where food waste is most often disposed of), as well as transport costs, operating costs at treatment plants and separation costs in some cases.

The production and disposal of waste in Europe generates 170 tonnes of CO2 emissions and uses 26 million tonnes of resources, with an associated cost of 143 billion euros.

To summarise, in addition to its economic and social impacts, food waste has consequences for the climate, water, land and biodiversity, which is why it is essential to take steps towards sustainable economic and social development.

Another concern that could arise from climate change has to do with volatility in food prices, a factor that could be a major source of food insecurity, as it will increase inequality in access to food. Changes in supply - due to the vulnerability of the climate - and demand - due to consumption trends - are decisive in determining the value of basic products such as corn and soya, for example.

These are more than enough reasons for us to reflect on and question this model of production and consumption, and for the sake of sustainable, balanced and fairer development, different modes of production, more responsible behaviour and more conscious consumption choices are crucial.

Food loss and waste is not just a matter of food security, it is a challenge in an increasingly globalised world whose sustainability ("living within limits") must be safeguarded for present and future generations.

[13] European Commission, Directorate-General for the Environment, Preparatory study on food waste across EU 27, 2010.
[14] Available at http://www.cienciaviva.pt/img/upload/do field to fork.pdf

CHAPTER 5

Causes of Food Waste

In addition to the contraction of the agricultural sector and the phenomena of urbanisation and globalisation, which have led to new lifestyles, there are various and diverse reasons given for food losses and waste throughout the various stages of the food chain (production, processing, distribution and consumption).

The report of the High Level Panel of Experts on Food Security and Nutrition "Food loss and waste in the context of sustainable food systems" (HLPE,2014)[15] , analyses the complexity and diversity of causes from a threefold perspective: (1) a systemic perspective, (2) a sustainability perspective, including the environmental, social and economic dimensions of sustainability and (3) a food security and nutrition perspective, considering how food loss and waste relate to the various dimensions of food security and nutrition, and concluded

> "Food losses and food waste are consequences of the way food systems work, technically, culturally and economically" (HLPE, 2014).

According to this report, food losses and food waste at each stage of the food chain can result from micro, meso and macro causes, all of which are interconnected, such as losses due to poor temperature management, a situation that can occur at any stage of the food chain.

Thus we have "micro-level" causes, which occur from production to consumption and which result from the actions (or non-actions) of individual actors, in response (or not) to external factors. Factors relating to poor scheduling and timing of harvests, harvesting techniques, poor harvesting practices, lack of adequate infrastructure, rough and careless handling of produce are major contributors to food waste.

In other cases, it is mainly issues of product standardisation that dictate their demise. The rigidity of standards, criteria and safety rules in the commercial supply chain (adoption of excessively tight expiry dates, highly valued aesthetic appearance of products, size and quality of packaging); poor stock management and oversupply on the part of distribution.

> "Some products are left unharvested due to an inability to meet certain quality standards (shape, size, weight) dictated by processors, retailers or target markets, contributing significantly to waste." (Stuart, 2009)

Likewise, intensive production models and the lengthening of the value chain, which increasingly distances the producer from the consumer, combined with the lack of good storage and refrigeration conditions, are factors that will lead to waste.

---

[15] HLPE (2014) http://www.fao.Org/3/a-i3901e.pdf

Like the usual distribution channels' preference for fruit and vegetables that are "perfect" in shape, colour and size, they are one of the main causes of waste.

As for consumption at home and in catering, and especially in more developed countries where food is used as a symbol of prosperity (IMechE, 2013), the propensity for food waste is high.

Behavioural causes are at the root of such waste.

- Habits of buying and cooking more than you need (promotions and advertising campaigns);
- poor preparation and consumption of excessive amounts of food; poor preservation of food in the home
- Poor food culture (not knowing how to reuse leftovers);
- lack of planning when shopping and lack of knowledge about food, which has led to its devaluation;
- lack of understanding of expiry dates ("use-by date" versus "minimum durability date");
- the restaurant industry's inability to forecast daily customers and the size of portions

are all examples of factors that contribute to food waste

Evans (2012) in his book "Beyond the throwaway society: ordinary domestic practice and a sociological approach to household food waste", illustrates some of the ways in which in a consumerist (and throwaway) society, the transition from "food" to "waste" arises as a consequence of the ways in which domestic practices are socially and materially organised.

The author pays special attention to households and the contingencies of everyday life, to social relations and how they manifest themselves and to the socio-temporal context of eating practices (more/less large families; families with/without children). At the same time, he points out that this concern about food waste will have to involve public and political concerns about the origins and consequences of food waste (Evans, 2012).

The "meso level" causes include secondary causes and structural causes.

Lack of equipment and/or good practices, inadequate organisation, coordination and communication between actors in the food chain (e.g. processing that renders the product unusable at a later stage in the chain, etc.), inadequate infrastructure, discredited economic conditions along the food chain (unmarketable product, etc.) are the main causes of waste along the food chain.

Retailers can also influence the activities of supply chains, dictating the quality of the product to be supplied and displayed in their outlets. Quality standards (in terms of shape, size and weight) imposed by processors, retailers or target markets can lead to the production and creation of food waste.

Finally, "macro-level" causes include those linked to policy and regulatory environments, as well as systemic causes, which can appear at various levels, such as a disorganised and uncooperative food system, the lack of institutional or policy conditions to facilitate the coordination of actors to enable investments and the adoption of good practices, the lack of regulations for the donation of surplus food from distribution or catering, etc.

To summarise, we have causes that can derive from a single actor, those that derive from a more collective action and those that can arise from the inaction (or inattention) of public authorities.

CHAPTER 6

The contribution of the circular economy to combating food waste. From waste to resource.

We live in a time marked by consumption (a structuring element of the most developed societies), where the search for the superfluous has led to waste and with (mis)thinking based on the idea that through solutions such as recycling everything would be solved.

However, the environmental and social unsustainability caused by a (linear) economy that extracts scarce resources and incorporates them into consumer objects, which after a short period of useful life are discarded in the form of waste, has been scientifically demonstrated, which is why a paradigm shift is essential. Growth and an economy that have been built on a model that is now under threat because of the limited availability of natural resources.

Hence the need for a resource-efficient economy that is restorative and regenerative in the way it is conceived, that preserves and respects natural capital and minimises environmental risks, an economy geared towards the conservation of natural resources.

To this end, and to ensure sustainable growth in the EU, in 2015 the European Commission adopted an ambitious new package on the circular economy, in which it makes the transition from a linear economy to a more circular economy essential, in which the value of products, materials and resources remains in the economy for as long as possible and the production of waste is reduced to a minimum[16] .

A new model where at the end of the process we can recycle the products, or part of them, and put them back on the market, i.e. "closing the loop" of the product's life cycle (Figure V), i.e. giving it a new chance, through more reuse, recovery and recycling, with great advantages both for the environment, by reducing the use of raw materials, and for the economy (from waste to resources).

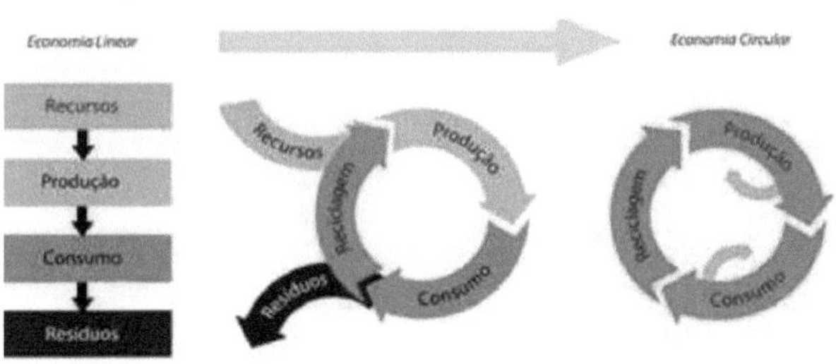

[16] **"Closing the loop - an EU action plan for the circular economy" COM/2015/0614 final, available at** https://eur-lex.europa.eu/legal-content/PT/ALL/?uri=CELEX%3A52015DC0614

Figure V - Objective of the Circular Economy: to build a society without waste, based on sustainable production and consumption practices (Circular Economy Portugal, 2017)

Due to the specific nature of its products or value chains and their environmental footprint, the European Commission has prioritised action, among other areas, to prevent/reduce food waste (and plastics).

To this end, he proposed that all Member States draw up national food waste prevention strategies and endeavour to ensure a reduction of at least 30% in food waste in the manufacturing, retail/distribution, hotel/restaurant and household sectors by 2025[17] .

[17] The year 2016 has been designated the European Year against Food Waste

CHAPTER 7

The food waste hierarchy

In this fight against food waste, the US Environmental Protection Agency has created a six-level Food Recovery Hierarchy[18] , which prioritises actions that citizens and organisations can take to combat food waste.

To this end, the European Union has proposed applying (adapting) the waste hierarchy to food waste, a hierarchy that prioritises measures to treat food waste.

[19]This led the European Parliament to create the food waste hierarchy (Figure VI), which emphasises the top three levels, as they relate to measures that can be taken before food becomes food waste and therefore offer more benefits for the environment, society and the economy.

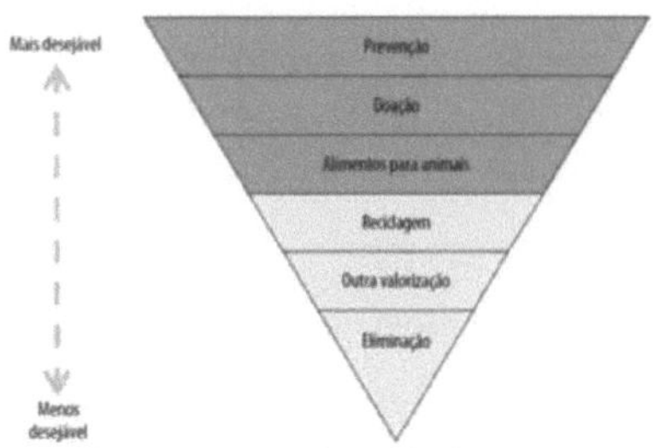

Figure VI - Hierarchy of food waste proposed by the European Parliament (ECA, 2016)

Each level of the food waste hierarchy focuses on different management strategies for your wasted food.

The first level has to do with waste prevention/reduction, from production to the end consumer, including raising awareness. To this end, new ways of producing and processing food should be adopted, from new crops to the resizing of packaging, as well as changing behaviour and eating habits, which can include less consumption of animal products and more legumes, and frequenting local markets and buying seasonal produce.

On an individual level, two people in particular have influenced the public agenda in the fight against food waste, publicly denouncing the environmental and social impacts of food waste: the British Tristan Stuart (founder of the Feedback movement) with his work "Waste: Uncovering

[18] Available at https://www.epa.gov/sustainable-management-food/food-recovery-hierarchy

[19] prevention, preparation for reuse, recycling, other types of recovery (e.g. energy recovery) and disposal.

the Global Food Scandal (2009) and with the "Feeding 5 Thousand" initiative[20] , and American journalist Jonathan Bloom, author in 2010 of "American Wasteland: How America Throws Away Nearly Half of its Food (and what we can do about it)", in which he tries to demonstrate how all of us, through conscious attitudes in our daily lives, can reduce and eliminate food waste.

In Europe, the "Zero Waste" movement (Refuse - Reduce - Reuse - Recycle - Rot) is a citizens' movement working towards zero waste (be it food, plastic, paper, etc.). Its main objective is to change the philosophy of waste management. Not just managing it by recycling and sending unused waste to landfills and incinerators, but eliminating all waste, in other words, being able to live without generating waste. To do this, we will have to change consumption behaviour and lifestyles, such as buying food in bulk or making compost from organic waste.

In the Netherlands, the government has pledged to reduce food waste by 20 per cent. To achieve this goal it followed the example of the successful initiative in the UK (the Love Food Hate Waste campaign) and launched the "taste the waste" campaign which teaches "How to save money without much extra effort".

In Britain, to combat food waste, supermarkets have opened that sell end-of-life items at symbolic prices.

In Denmark, the NGO WeFood organised a crowdfunding campaign that made it possible to open a surplus food supermarket that sells half-price or less food that is expired or has damaged packaging, presenting itself as a low-cost alternative for families on low incomes (The Uniplanet, 2016).

Finally, because we live in a networked society, I'd just like to take note of some innovative solutions for combating food waste that come from the technological field, which allow any donor to make the type and location of food ready to be donated available on their mobile phone app.

This is the case with restaurants that post meals online that have not been sold at lunch/dinner times and can be purchased later at reduced prices ("To Good to Go" in several countries; "ResQ Club" in Finland; "Gebni" in the United States) or at the consumer level, the OLIO app - The Food Sharing Revolution, a free app in the UK that connects people with neighbours and local shops so that surplus food in the home and community can be shared and not wasted.

The 2nd level of the hierarchy is donation[21] , the reuse of food that would otherwise go to

[20] The initiative consists of organising big lunches with fresh food that would otherwise go to waste.

[21] Donating surplus food from food sector activities to people in situations of social vulnerability is an

waste, whether from cafés/restaurants or large commercial supermarkets. By donating this surplus, we feed people and support local communities while saving money and resources.

Although it is still a process with many obstacles, which put off potential donors (the question of responsibility, food safety, the lack of tax benefits), there are already good initiatives in Europe to donate.

In France, a law was passed in 2016 that obliges large retailers and food distributors with an area of more than 400 square metres to donate their surplus (unsold or expired products) to food charities that are still in edible condition.

In Italy, a bill against food waste has been approved by the government, which provides for awareness-raising and incentives for restaurant and supermarket owners to give away their excess food. The Italian state is even proposing to reduce the tax rate on rubbish in proportion to the amount of food donated and to cut down on the bureaucracy associated with the food donation process.

In Portugal, the Large Distribution sector has set a great example by fully assuming its social responsibility. In addition to **raising awareness of good practices among customers on a daily basis, they regularly** donate food (end-of-life products) to charitable organisations, always safeguarding quality, hygiene and food safety.

The third level is the reuse of waste for animal consumption: food that is not suitable for human consumption can be transformed into animal products. This is a factor that may lead us to reflect on the fact that the human supply chain and the animal supply chain are not completely independent and animal production is ultimately intended for human consumption (Batista, et al, 2012), as is the case with fruit that is not of sufficient quality for the market and is often destined for animal feed.

The 4th level is industrial recycling, through options that aim to alleviate some of the environmental and economic issues associated with food waste, while increasing the use of alternative energy sources. Around 40% of what we throw away is bio-waste that can be valorised to produce energy (biogas, biodiesel).

The 5th level of the hierarchy is composting (individual or communal), a process that transforms organic matter (food waste) into organic fertiliser in order to improve soils, grow subsequent crops and improve water quality.

---

effective way of combating hunger, eradicating poverty and reducing inequalities, three global goals for sustainable development (SDGs 1, 2 and 10).

Finally, we have the 6th level, which corresponds to waste that cannot be recovered in any way, leaving no option but to incinerate it or dump it in landfill sites.

CHAPTER 8

Combating food waste on the global agenda

The FAO, after launching the "Zero Hunger Challenge" in 2012 in order to achieve total sustainability of food systems, has implemented a global initiative "SAVE FOOD: Global Initiative on Food Loss and Waste Reduction" which aims to encourage dialogue between the various stakeholders, industry, politics, academia and civil society in order to create solutions to combat food waste. With this in mind, the FAO plans to calculate a food loss index for each country and a global aggregate.

In the US, ReFED " Rethink Food Waste Through Economics and Data" a non-profit multi-stakeholder organisation with an influential network of the country's leading business, non-profit, foundation and government leaders, is committed to scaling up solutions to reduce food waste, particularly at household level.

The World Resources Institute (WRI), a Washington-based environmental non-governmental organisation that works with institutions and partners around the world to develop and promote policies aimed at protecting the Earth and improving people's lives, has been focusing on food waste. To this end, in 2016 it created a quantification protocol (FLW Protocol) to count and report food loss and waste on a global scale[22] .

The OECD (Organisation for Economic Co-operation and Development) is also developing a line of work dedicated to Food Waste. The "Food Chain Analysis Network" was launched in 2010 by the Department of Trade and Agriculture to debate topics related to the functioning, sustainability and productivity of the agri-food system[23] . Under the theme "Reducing food loss and waste in the retail and processing sectors", the OECD concluded that the lack of a commonly agreed definition is a limitation to the progress of the work, identifying the need to improve the coverage of information for the food supply chain.

In the European Union, we have had the Fusions project since 2012, which aims to make Europe more resource-efficient by significantly reducing food waste.

The growing waste produced in Europe, particularly in urban areas where the vast majority

---

[22] The Food Loss & Waste Protocol (FLW Protocol) available at https://www.wri.org/our-work/project/food-loss-waste-protocol

[23] It should be noted that agriculture contributes to climate change and is affected by it. At each stage of the chain, greenhouse gases are emitted into the atmosphere, especially in livestock farming. In turn, climate change (temperature, rainfall) affects crops.

of the world's population is expected to live by 2050, represents a cost to society and a serious problem for the planet.

In 2015, on the occasion of the Milan Expo "Feeding the Planet, Energy for Life", the "Milan Pact on Urban Food Policy" was signed, a document whose purpose is to involve as many cities and local authorities around the world as possible in the development of food systems based on the principles of sustainability and social justice.

With the aim of improving the environment in urban and peri-urban areas, since 2016 we have also had the FORCE project "Cities Cooperating for Circular Economy"[24] (where Portugal is present through, among others, Lipor and the Dariacordar association) funded by the EU, which aims to discuss innovative processes and solutions with sustainable economic models that contribute to social change and the reduction and reuse of waste.

In the United Kingdom, the work of WRAP (Waste and Resources Action Programme), a non-profit organisation that works with governments, companies and communities to propose solutions and improve resource efficiency, has stood out, particularly with actions to combat food waste, such as the "Love Hate Waste" campaign launched in 2007.

---

[24] Available at http://cordis.europa.eu/project/rcn/207269 en.html

CHAPTER 9

Combating food waste in Portugal

In Portugal, the National Commission to Combat Food Waste (CNCDA), created in 2016, has the mission of promoting the reduction of food waste through an integrated and multidisciplinary approach. To this end, it drew up the National Strategy to Combat Food Waste (ENCDA) and the Action Plan to Combat Food Waste (PACDA). Identifying good practices at national and international level in the fight against food waste, as well as promoting the involvement of civil society organisations, are among its main measures.

At national level, there are various initiatives to prevent and reuse food waste that occurs throughout the food supply chain, from awareness-raising campaigns to the collection from farms of products rejected by the commercial circuit or the redemption of surplus products from department stores.

In catering, we have good initiatives such as "Dose Certa", an innovative practice that resizes the size of the portion to be served in the restaurant, and the "Embrulha." project (Lipor, 2016), which encourages restaurants to invite each customer to take home what was left on the platter. (Lipor, 2016) which encourages restaurants to invite each customer to take home what was left on the platter, thus breaking a stigma from the past.

The Large Distribution sector has also committed itself to actions and initiatives to reduce the social and environmental consequences of food waste.

To this end, there has been a strong effort to raise awareness among consumers of the need to adopt more conscious and sustainable purchasing, preservation, cooking, storage and consumption habits, such as the appeal to consume "ugly vegetables" (non-standardised), the sale of loose products, the use of smaller packaging and the placing of an "orange label" on products whose expiry date is nearing.

We also give the example of "GoodAfter", the first Portuguese online supermarket to combat food waste. This is a Portuguese platform that sells food and non-food items that are close to their preferred consumption date, but "with a total guarantee of food safety and can therefore be sold at more favourable prices". The goods can be purchased with savings of up to 70% compared to the prices charged in traditional supermarkets.

In this article we will try to understand how in Portugal two social entrepreneurship projects from civil society - the Fruta Feia co-operative and the Re-Food association - are contributing to the reduction of food waste in a sustainable way.

CHAPTER 10

The "Fruta Feia" co-operative: A case of social and useful entrepreneurship

Since the end of the 20th century, driven by the transition to the information and knowledge society, a new wave of entrepreneurs has been helping to revitalise the economy.

When Peter Drucker discusses the "entrepreneurial society" within the spirit of contemporary capitalism, he defines it as an important turning point in history, a kind of revolution "without bloodshed, civil war, or concentration camps, without economic catastrophe, deliberately, with direction, and under control" (Drucker, 2011).

The Austrian economist Joseph Schumpeter, in his 1942 work "Capitalism, Socialism and Democracy", describes the entrepreneur as one of the most important forces in capitalism, whose evolution corresponds to cyclical stages of expansion and crisis. For Schumpeter

> **"The function of the entrepreneur is to reform or revolutionise the production model,** thus participating in the **process of "creative destruction" of the prevailing economic order.** He **is** therefore responsible for **innovation and for the economy's ability to develop." (Schumpeter, 1985)**

Similarly, entrepreneurial behaviour requires a willingness to take calculated risks and the possibility of failure. In short, the entrepreneur aims to develop and implement sustainable solutions to neglected problems in society. Someone who looks for opportunities, takes risks and innovates.

Sustainability is of great importance in environmental issues, as it is a way of securing natural resources for present and future generations. A conscious entrepreneur with sustainable and innovative practices creates value for society and involves and influences the people around them.

> Social entrepreneurs are agents of change in society, creators of innovations that disrupt the status quo and transform our world. By identifying the people and programmes that are already bringing about positive change, we enable them to extend their reach, deepen their impact and fundamentally improve society (Skoll Wordl Forum, 2014).

In Portugal since 2013, as part of the European project "FLAW4LIFE - Spreading Ugly Fruit Against Food Waste"[25] , we have the consumer co-operative "Fruta Feia", whose motto is "Beautiful people eat ugly fruit". It's a social entrepreneurship project created by a young environmentalist who, by exploiting a market inefficiency, seeks to "disrupt the status quo and transform the world", i.e. work for social change.

---

[25] Available at http://www.flaw4life.com/pt

An innovative project against food waste due to appearance, which at the same time aims to promote responsible, local and sustainable consumption.

An alternative market which, by reincorporating into the consumer chain fruit and vegetables that would otherwise end up in the rubbish bin simply because they don't meet the aesthetic standards required in distribution, not only saves the resources associated with producing this food, but also generates value for farmers and consumers.

> **"If at the beginning it was an idea, a desire, there was an associated risk, we didn't know if it was going to work.** Saving more than a thousand tonnes of vegetables in four and a half years (15 tonnes a week), which would otherwise have gone to waste due to their appearance, having 162 farmers as partners, reaching 11 employees, 11 delivery points and 4,600 members (with 14,000 on the waiting list), means that we're not just an idea, it's a model that works, a consumption proposal that many people agree with and that has proven its sustainability, both socially and environmentally, and environmental, as we follow a policy of proximity consumption (we don't travel more than 70 kilometres from the delivery points) and financial, as the money we earn is used to pay operating costs, a fair price for farmers for their products and a fair wage for workers (excerpt from an interview given by Isabel Soares in December 2017).

Producers have the opportunity to sell products that were rejected by distributors for reasons of appearance and size alone. For their part, consumers (who don't judge quality by appearance) can buy products from their region at a lower price.

A case of clear economic rationality that only benefits the local economy, with a greater social dimension and respect for human beings and the planet.

Fruta Feia's main objectives are:

> "reduce the tonnes of quality food that are returned to the land every year by farmers (regardless of whether they are produced organically, in integrated or conventional production) and thereby also **avoid** unnecessary waste of the resources used in their production, such as water, arable land, energy and working time, as well as reducing GHG emissions resulting from agricultural production; change consumption patterns, so that in the future all quality fruit and vegetable products are marketed equally, regardless of size, colour and shape; raise public awareness of the problem of food waste and the fact that ugly food is not rubbish, while demonstrating the possibility of consuming seasonal and regional products at a lower price" (Fruta Feia, 2017).

Every day, Fruta Feia goes out to the countryside to collect all the produce that hasn't made it into the commercial circuit. During the afternoon, this produce is sold at reduced prices to its beneficiaries, who can choose between small (three/four kilos and five to seven varieties) or large (six/eight kilos and seven to nine varieties) 'fruta feia' baskets. The baskets are made up of seasonal and regional products, which vary from week to week according to the time of year.

CHAPTER 11

The Re-Food Movement. A movement by and for the community

When there is surplus food, the best course of action is to redistribute it for human consumption. Redistribution consists of a process in which surplus food that might otherwise be wasted is recovered through donation, collected and supplied to citizens, especially those most in need, always in strict compliance with the rules of food traceability[26] and food hygiene and safety standards (EC, 2015).

In this article we have analysed the Re-Food 4 Good Association, currently a Private Social Solidarity Institution, whose motto is "Making the most of food" and which, in the context of environmental and social responsibility, fights food waste and hunger in local communities, while strengthening local social ties and enabling adherent partners to actively exercise their social responsibility.

An authentic case of a social, solidarity and inclusive economy that proposes a change in dominant thinking capable of transforming current economic relations into an approach based on co-operation, citizenship, solidarity, local development and sustainability.

"We all have the power to change the world, each in our own way" (Interview with Hunter Halder,2017)

This is an independent organisation founded by Hunter Halder, a social entrepreneur, who has set up a micro-local action project (the area of operation should be approximately 2 square kilometres), with its own model (Figure VII) with the aim of reusing surplus food from restaurants and cafés, which would otherwise end up in the rubbish or landfill every day. It's a unique movement that works in a different way to other food rescue operations and in a different way to other social movements.

[26] Traceability is the ability to know the history, use or location of a foodstuff or an activity through recorded means of identification. In practice, it means being able to locate and identify the product at each of its stages. At Re-food 4 Good, all the food rescued is immediately labelled by the volunteer at the time of collection, and then at the operations centre all the steps are recorded by the managers: the donating partner, the collection volunteer, the day (colour label) and time of collection and the beneficiary to whom the meal was distributed.

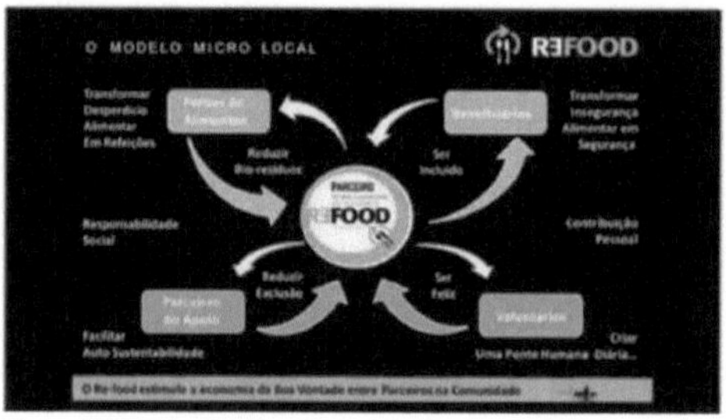

Figure VII - The Re-Food Model (Re-Food 2016)

An idea launched in 2011, at a time of serious economic and financial crisis in Europe and Portugal, a situation that caused major imbalances in society (unemployment and job insecurity) and a profound change in people's social lives (severe austerity)[27].

On top of all this, the government's retreat from social policies was reason enough to throw a significant part of the population into dependency on state subsidies and social exclusion.

Aware of the serious food shortages suffered by friends in his community (the parish of Nossa Senhora de Fátima in Lisbon), Hunter Halder couldn't remain indifferent to the food **waste he saw every day at closing time in restaurants and cafés in the city of Lisbon.** Food that was in perfect condition **but which, having lost its commercial value at the end of the day, was thrown in the rubbish bin every day. According to Hunter,** ***"All food has its economic value until closing time of the establishment (...) after this moment - if the food - it becomes waste, even though its nutritional value is still intact and it is therefore in good condition for human consumption. This is where Re-Food comes in, because by collecting this food on a daily basis, it won't allow this surplus to go to waste." (Interview with Hunter Halder in 2017).***

Throwing away perfectly good food is an act of exclusion. In fact, what is left over for many is left lacking for those who have nothing. The dignity of every human being appeals to everyone's ethical conscience (individual and collective), in the sense of being attentive to the needs of others and not living exclusively centred on the abundance of goods one enjoys.

"We have made the fruits of the earth a commodity for some, thus generating exclusion. Consumerism has led us to become accustomed to the superfluous and the daily waste of food, which we are sometimes no longer able to give the right value to and which goes beyond mere economic parameters. We would do well to remember that food that is wasted is like food that has been stolen from the table of the poor, the hungry." (Address by Pope Francis

---

[27] **"International financial assistance programmes are recognised for their lack of social concern and their impact on the social determinants of health, particularly their impact on poverty, the main determinant of health" (Marmot, 2005).**

at the headquarters of the World Food Programme (WFP) Monday 13 June 2016) This is an association, supported 100% by citizens, which does not fundraise, does not own property and does not own anything that does not serve its mission. Its action in the community is geared towards inclusion and welcoming, where everyone is invited to participate as they see fit and no one is excluded, as Hunter says: "***Local institutions (grocery stores, pharmacies, etc.) are often an invaluable help in the sense that they know the people in the neighbourhood and have information about who needs food (shame hunger), so I always ask all our coordinators and managers to invite everyone in our community. They can join us as volunteers or by -matematising -who needs -help (...) - so they are given the opportunity to exercise their duty of social responsibility in their community through the Re-Food project" (Interview with Hunter Haher in 2017).***

For Hunter, the greatest resource identified and activated by Re-Food is not the abundance of food, the physical spaces or the support of local businesses, but the goodwill of people in every community who voluntarily come together to build the human bridge between excess and need that effectively changes the world for the better in their own neighbourhoods.

At the same time, Re-Food prevents many tonnes of food from spoiling and going to landfill, reducing the amount of waste produced. In this way, Re-Food provides an exemplary work of economic and environmental sustainability and social inclusion: ***because the food, labour and facilities are provided*** free of charge and the unavoidable operating costs are very low; because each meal rescued reduces the negative impact of biomass degrading in landfills and most of the collection work is carried out by teams on foot, by bicycle or with donated electric vehicles; because it has the ability to awaken civic-mindedness in its ***participants and to include the entire community*** (Hunter Halder, 2016).

"(...) it gives us immense pleasure to feel how much the Portuguese recognise and are receptive to joining this project. We are delighted to see that the Portuguese, from north to south, both individually and institutionally, are aware of the absurdity of throwing away surplus food when there is hunger, and actively want to fight hunger and waste by offering their time and talent to Re-Food" (Hunter Halder, 2017).

Its vision is "the creation of a new world where: everyone has the food they need, all food is utilised, all citizens actively participate in the management of precious community resources and everyone assumes their power, their right and their obligation to make the world a better place" (Re-Food website, 2016).

A low-cost, high-productivity project based on values such as equality, respect, social inclusion, sustainability and optimism.

An organisation that works on the basis of volunteering and partnerships and whose mission is to eliminate food waste and end hunger, including all members of the community in this effort.

Local institutions, both public and private, are given the opportunity to provide the (often disused) physical space or any other kind of support available to facilitate the running of Re-food in their locality, while local businesses are invited to participate in the community effort in the way that best matches their social responsibility goals - the whole community is invited.

In fact, all companies have a duty of social responsibility, but not all companies find the opportunity to fulfil this duty.

A movement that wants to be of everyone and for everyone, regardless of creeds and beliefs (political or religious) where citizens will always dictate the rules," emphasised Hunter Halder.

> "What ***impressed me most was Re-Food's ability to integrate very different people without any kind of prejudice... and that's not common nowadays. (...) All these people spend time together here and build a positive atmosphere.***
>
> ***In fact, Re-Food provides us with a great personal experience and a great life lesson." Testimonial from 1 Re-Food volunteer***

The Re-food Movement's actions have endured over time largely due to internal mechanisms that guarantee its economic, environmental and social sustainability.

All the food is freely donated by the partners (food sources), all the work is carried out by volunteers motivated by goodwill, the operating space is donated free of charge and the unavoidable operating expenses are very low - an average of around €250 per month (easily financed by the goodwill of the community at large). The thousands of meals that are rescued and delivered each month by each Re-food centre are produced for less than 10 cents each (and are, of course, delivered free of charge). Economic sustainability thus flows from the effectiveness of the model and the goodwill of the community.

On the other hand, the Re-food model makes it possible to maximise the collection of food, both when minimising the ecological footprint (collection work carried out on foot or by bicycle) and in more distant collections with a low environmental impact (donated electric vehicles). Likewise, each meal rescued reduces the negative impact of the biomass that degrades the environment in landfill sites (Figure VIII).

Figure VIII- Re-Food Operating Model

Their innovative processes allow them to work together to solve the most challenging problem of food rescue in a commercial environment: "how to provide a sustainable food distribution service to many, different and dispersed collection points every day - and in real time"

The tangible social impact is proven by the positive effects on volunteers and the community in general. A simple and inclusive organisational model that, by awakening civic spirit in its participants and the entire community, guarantees human sustainability and has contributed to the exponential growth of the Re-food Movement (Figure IX).

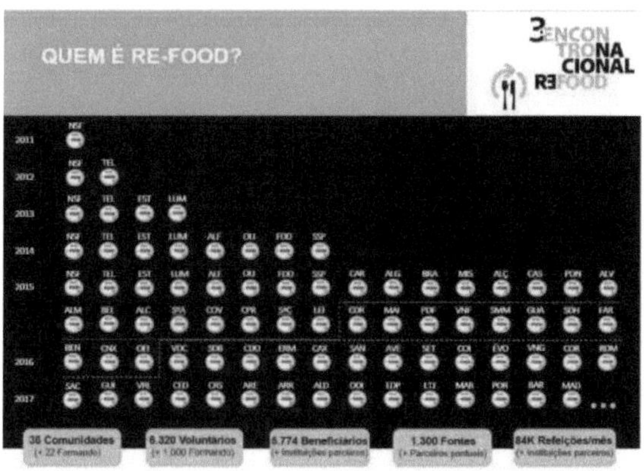

Figure IX - Re-Food Movement in 2017 (updated) [28]

In order to achieve its goals, Re-food relies on a team of volunteers, food partners (restaurants, hotels), partnerships with public (schools) and private institutions (department stores or events such as "Rock In Rio, Lisbon" or "Optimus Alive") and financial support from national partners, and is also open to donations.

Every evening at closing time, when everyday products lose their economic value, several

[28] A project that has grown exponentially since then (from 1 centre and 1 volunteer in Lisbon in 2011 to 40 centres and 6300 volunteers across the country in 2017). In 2017, nearly 1 million meals a year (84,000/month) were rescued from 1,300 food sources, assisting 5,700 beneficiaries (1,300 families), which represents an annual reduction of 500 tonnes of bio-waste (Re-food, 2017).

teams of volunteers take to the streets to rescue surplus food from cafés and restaurants and immediately redistribute it to people in need of food.

- Volunteers

This is a very important sector for Re-Food, whose mission is to ensure that volunteers are welcomed, cherished and incorporated into the different teams, depending on their availability and motivation, so that they feel included in the project.

"Volunteering is the set of actions of social and community interest carried out in a disinterested manner by people, within the scope of projects, programmes and other forms of intervention at the service of individuals, families and the community developed on a non-profit basis by public or **private** entities. **(...) The volunteer is the individual who freely, disinterestedly and responsibly** undertakes, according to their own aptitudes and in their free time, to carry out voluntary actions within the scope of a **promoting organisation.**"[29] .

Social volunteering is one of the ways of organising those who want to put their knowledge, time and determination at the service of reducing unequal opportunities and other barriers that limit the citizenship of people in disadvantaged situations.

It applies to all those who give their time without expecting a monetary reward. They have different skills, competences, backgrounds and motivations. It is up to the organisation to define the necessary procedures to ensure that they have the best possible experience (KnowHowNonProfit, 2016).

In Portugal, in terms of citizenship and participation, we have a fragile civil society, with a very low level of citizen involvement compared to other European countries (there are many associations, but few people involved), which is not good for democracy, as it will be better sustained the more people actively participate in it.

The Portuguese are among the Europeans who do the least voluntary work. In 2012, according to the INE, one million people, or 12% of the population over the age of 15, had taken part in voluntary work at least once, a far cry from the rates achieved in Northern Europe, especially in the Netherlands, with 57% of the resident population aged 15 and over claiming to do voluntary work. Portugal's position is explained, "in part, by the culture of participation in voluntary work activities and by the country's socio-economic conditions, with some correlation being observed between the degree of economic development and the rate of volunteering in countries" (Instituto Nacional de Estatística, 2012).

Citizen participation fosters interpersonal trust and tolerance, social well-being, personal and social skills,

[29] Bases of the legal framework for volunteering (art. 2 and 3 of DR No. 254 , Series I-A 3 November 1998 3 November 1998)

civic virtues, psychological empowerment and a sense of belonging and community (Putman, 2001; Sullivan and Transue, 1999), which can have a social impact by increasing civic culture, social capital, the climate for deliberation and discussion and pluralism (Putman, 2001).

People who work in third sector organisations (organisations that come from civil society) are above all attracted by the moral and ethical values and objectives that these organisations hold dear. They want to know and feel that their time is well spent, organised and that the contribution they make to the institution is valued.

Men and women, young and old, students and pensioners, employed and unemployed, all with different skills, competences, backgrounds and different motivations for volunteering, come to Re-food to offer small amounts of their most precious resource - time - to change the world around them. "Their gift is so great that, in giving, they also receive." (Re-Food, 2016)

"Goodwill is always huge - it is," emphasises Hunter, "Re-Food's greatest asset", but you have to *BE* realistic. ***"Many people have an idealised idea of volunteering. Volunteering is work.*** "That's why some people come and quickly give up. ***"We have to have In*** szs/ewflyZexzve/ *AND* ízceztar the people? cozwo ***elac*** are. " He adds, ***"The vohintarians are the first beneficiaries of the work they do - they feed themselves by feeding the*** owirs." (Interview with Hunter Halder in 2017)

- Beneficiaries

It is Re-Food's responsibility to identify and guarantee the well-being of the entire beneficiary population (individuals, nationals and foreigners, families and children, homeless people, etc). All those who need food support, temporarily or in the long term, often have other needs too. To do this, you have to be close to them, get to know them and talk to them, understand their anxieties and concerns, and in many cases you can and should refer them to other institutions that can provide them with other types of support (parish council, social worker, health centre, employment centre, other associations).

It's a process that results both from the involvement of local institutions (church, parish council), which know those who need food in the local micro-area, and from the use of indirect techniques to identify people in need who are not supported by any other institution from other agents in the local community (pharmacies, grocery stores).

It should be remembered that in Portugal, according to INE data, in 2016 19.5 per cent of the population was in a situation of material deprivation and 8.4 per cent in a situation of severe material deprivation.

In this context, Re-Food "helps people struggling with food insecurity in local communities. Priority is given to the most vulnerable members of the community - children, the elderly, the disabled - but anyone who needs food is welcome to receive support from Re-Food." (Re-Food, 2016)

-it was a way of doing something for others...I finish school early - there's plenty of time left over in .... Holland... my friends and I already do voluntary work...here I've done all 3...

tasks and the one I identify with the most is distribution...talking to those who need it, listening to their complaints and giving them advice...I like that...it has something to do with me...there are old people there who don't have anyone to talk to during the day...so have you noticed that they don't mind arriving here very early (at 6pm...) and then they don't leave until 9.30pm...I'm convinced that this is the best time of the day for these people...at the same time it's sad.... " testimonial from 1 Dutch Erasmus student

- Partners

By affirming their social responsibilities towards the community and assuming

By voluntarily making commitments that go beyond the fulfilment of all legal obligations, the partners are giving a clear sign of their commitment to development based on the three pillars of sustainable development: environmental, social and economic.

"There are many restaurants that, fearing that they won't be able to fulfil their daily commitment to Re-Food, set aside some food for this purpose during the day. There are others who, in cases where there is nothing left at the end of the day, deliberately cook something themselves (soup or a dish) in order to deliver it to Re-Food." **(Hunter Halder, 2017)**

Being recognised across borders is the current major objective ("parish, city, planet"), despite already having branches in Madride in several cities in India (The Robin Hood Army), with others in talks (in Amsterdam, Milan, Budapest, Brussels, London, Paris, Rio de Janeiro, Luanda and Maputo).

To conclude, we can say that Re-Food gives meaning to the community, involving and including everyone, through a common goal, which is the elimination of the "waste of life".

It creates and strengthens its social capital, enables its volunteers to freely exercise active citizenship and solidarity, and allows all partners to carry out good social responsibility practices in order to promote the good environment and greater social justice.

CHAPTER 12

Discussion and Limitations

To talk about sustainable development is to talk about the paradigm of today's world. The word development, which used to be synonymous with progress and growth, now has a different meaning: the sustainability of the planet, based on environmental, social and economic premises. In a modern, global society where a culture of consumption and waste prevails (consumption is one of the main forms of identity production, social reproduction and differentiation in society), we are witnessing the devaluation of the real value of food, which is now just another commodity and no longer a social good that should be accessible to everyone.

In the short term, climate change cannot be avoided, so mitigation and adaptation actions must be developed.

Throwing away healthy, edible food is not only immoral and unjust, but it also has social, health, environmental and economic impacts that we can only be concerned about. We cannot tolerate cases of hunger when there is food for everyone. Our planet has enough resources to feed the entire world population.

We therefore need to produce more with fewer natural resources, while keeping the ecological footprint of our food within sustainable limits.

The fact is that, in addition to the useless expenditure of environmental and economic resources associated with any form of waste, in the case of food waste we are also challenged from a moral point of view by the fact that millions of tonnes of food are thrown away every year, in a world where one sixth of the world's population goes hungry.

The first step towards a more sustainable resolution of food waste will be to adopt sustainable production and consumption (SDG 12.3) and address food surplus and waste throughout the food supply chain, rather than focusing solely on consumption. Food losses and waste represent an important picture of the inefficiency of our food systems.

It is therefore important to reflect on the way we relate to the environment, which will require greater rationalisation of the entire production and consumption model and greater social responsibility on the part of all of us (using local markets, valuing local and national production, eating seasonal fruit, eating less meat and more vegetables, etc.).

The challenge is enormous, but so are the collective resources we have in our possession.

Despite the great uncertainty about the data on food waste, the amount of food produced that is wasted seems to be high, which is why "closing the loop" on food is essential, making the prevention of food waste a priority for the European Commission as part of the new circular economy package.

By reintegrating (food) waste into the economy as a resource (reuse, recovery and recycling) and maximising its efficiency throughout the value chain, the cycle is closed: nothing is lost, everything is transformed in a continuous, nature-friendly reuse.

In this sense, the food waste hierarchy, by considering the three dimensions of sustainability (environmental, economic and social), offers a holistic and more sustainable approach to tackling the issue of food waste, and where the emphasis is on prevention, since it is more beneficial to avoid waste than to have to fight it later.

For an economically sustainable society, we need new ways of thinking, more supportive and inclusive local communities, a healthier human culture and collaborative, entrepreneurial social solutions for social change.

In addition to the need for greater environmental awareness, citizens' understanding of food, food safety and food waste and its causes must be improved. The industrialisation of food has given rise to the idea that we know less and less about the food we eat.

The lack of knowledge and "connection" to food, especially among the younger generations, as well as ingrained eating practices are some of the causes pointed to for the sharp rise in food waste.

At a time when concerns are growing about the effects of climate change, it is highly advisable to find alternative paths and transform communities into sustainable models that are less dependent on external resources, more connected to nature and more resilient to external crises, both economic and environmental (e.g. urban gardens, family farming, fair trade, the slow food movement, short supply circuits, etc.).

Local actions and initiatives that manage to collaboratively find solutions to create communities that are more focused on solidarity, group work, open and inclusive communication and respect for everyone's ideas, which increases collective social awareness and a sense of belonging to the community, so innovative attitudes and the ability to take risks are essential factors in promoting sustainable development.

In this sense, we value the efforts of social and sustainable entrepreneurs who, by seeking solutions to the most varied social and environmental problems, present themselves as active agents and transformers of society's values and behaviours, as the example of the Fruta Feia cooperative and the Re-Food movement shows us so well.

Only if citizens are more informed (and nonconformist), more aware of themselves and their surroundings (inequalities and injustices, violations of human rights, food waste) and willing to intervene in public life (either to show solidarity with people or causes, or to exert political and economic pressure on governments and companies to better manage resources) will we be able to aspire to a more democratic, fairer and less unequal society.

Hence the extreme importance that education (education for citizenship and environmental education) can play in building a critical and reflective collective conscience, in the sense of effective, participatory citizenship that has the capacity to respond to social problems and where we can reverse the typical "wait and see" attitude (not acting because others don't).

In addition to the difficulty experienced in the aforementioned lack of a uniform definition for the concept of food waste (food items are produced, processed and consumed in very different ways in different parts of the world), the limitations we encountered include the general misinterpretation of the term "waste".

In fact, the terms/expressions institutionalised in the public and political debate on combating food waste, such as "WASTE"; "SURPLUS"; "REST" are loaded with negative and pejorative meaning in society, so it would be advisable to replace them with others, perhaps "REUSE", although we are aware that not all waste can be reused. A situation to review.

In conclusion, and for future work on the problem of food waste, here's a thought.

The success of the response to food waste will depend on a holistic, multi- and interdisciplinary approach (economics, biology, human ecology, history, geography, anthropology, sociology, nutrition sciences and new information technologies) and across sectors (primary production and agro-industry, distribution, catering, end consumers), with the participation of political power and the involvement of civil society, and which integrates educational and environmental concerns and the fight against poverty and social exclusion. Only an integrated approach will lead us to understand the phenomenon of food waste, which is undoubtedly one of the most pressing challenges facing society today, and to which we urgently need to respond.

References

Bauman, Z. 1999. Globalisation: the human consequences, translated by Marcus Penchel. Rio de Janeiro: Jorge Zahar Ed.

_________. 2004. Liquid Modernity. 1. ed. Rio de Janeiro: Jorge Zahar.

_________ 2005 Liquid life. Translated by Carlos Alberto Medeiros. Rio de Janeiro: Jorge

Zahar.

Baptista, P., Inês Campos, Iva Pires, Sofia Vaz. 2012. From Field to Fork, Food Waste in Portugal, Lisbon: CESTRAS

Baudrillard, J. 2008. The consumer society. Lisbon: Edições 70.

Castells, M.2007 The Network Society; The Information Age: Economy, Society and Culture, Calouste Gulbenkian Foundation,4ª edition

Encyclical Letter Laudato Si' of the Holy Father Francis on Care for Our Common Home. Vatican City, Rome, 24 May 2015. Available at http://w2.vatican.va/content/francesco/pt/encyclicals/documents/papa- francesco 20150524 enciclica-laudato-si.html

EC, EUROPEAN COMMISSION.(2015) "Closing the loop - An EU action plan for the Circular **Economy". 2015. Available at** https://eur-lex.europa.eu/legal-content/EN/TXT/?uri=CELEX%3A52015DC0614

Contreras, J. 2011 Food modernity: between overabundance and insecurity. UFPR edition. 54. ed. Curitiba

European Commission.2017. "Towards a circular economy". Environment for Europeans magazine. Available at https://ec.europa.eu/environment/efe/all-articles en Cooperativa Fruta Feia. 2017. Available at https://frutafeia.pt/

Drucker, P. F. 2011 Innovation and entrepreneurship: practice and principles. São Paulo: Cengage Learning.

Durkheim, E. 1989, As Regras do Método Sociológico, Editorial Presença, Lisbon (1st ed., 1895).

EU FUSIONS, (Food Use for Social Innovation by Optimising Waste Prevention Strategies). Estimates of European food waste levels. Reducing food waste through social innovation. 2016. Available at : https://www.eu-fusions.org/phocadownload/Publications/Estimates%20of%20European%20food%20waste%20levels.pdf

EUROSTAT. 2017. "Income distribution statistics". Accessed at http://ec.europa.eu/eurostat/statisticsexplained/index.php/Income_distribution_statistics/pt Evans, D., 2012. "Beyond the throwaway society: ordinary domestic practice and a sociological approach to household food waste". Available at https://pt.scribd.com/document/334286929/EVANS-Beyond-the-Throwaway-Society

FAO. 2009. Food and Agriculture Organisation of the United Nations. "Proceedings of the Expert Meeting on "How to feed the world in 2050". Rome.

FAO,WFP, & IFAD. 2012. "The State of Food Insecurity in the World 2012. Economic growth is necessary but not sufficient to accelerate reduction of hunger and malnutrition". Rome: Food and Agriculture Organisation of the United Nations. Available at http://www.fao.org/docrep/016/i3027e/i3027e.pdf

FAO, IFAD, UNICEF, WFP, & WHO. 2017. The state of food security and nutrition in the world 2017: Building resilience for peace and food security. Rome: Food and Agriculture Organisation of the United Nations. Available at http://www.fao.org/3/a-I7695e.pdf.

Fusions. 2016. Food Use for Social Innovation by Optimising Waste Prevention Strategies. "Estimates of European food waste levels". Available at https://www.eu-fusions.org/phocadownload/Publications/Estimates%20of%20European%20food%20 waste%20levels.pdf

Giddens, A. 1991. The consequences of modernity. Translated by Raul Fiker. São Paulo: Editora UNESP

Guyomard, H., Darcy-Vrillon, B., Esnouf, C., Marin, M., Momot, A., Russel, M. & Guillou, M. 2011. Eating patterns and food systems: critical knowledge requirements for policy design and implementation. INRA. Document prepared for the Commission on Sustainable Agriculture and Climate Change . https://ccafs.cgiar.org/sites/default/files/assets/docs/guyomard et al eating patterns and food systems.pdf

Gustavsson, Jenny, Christel Cederberg, Ulf Sonesson, Robert van Otterdijk and Alexandre Meybeck. 2011. "Global food losses and food waste". Rome: FAO.

Harvey, David .1989. "The compression of time-space and the post-modern condition". In: The postmodern condition. São Paulo: Loyola

IMechE.2013. Institution of Mechanical Engineers. "Global food waste not, want not". Accessed on http://www.imeche.org/docs/defaultsource/reports/Global Food Report.pdf?sfvrsn=0

KnowHowNonProfit, (2016). Volunteers and your Organisation. Available at: https://knowhownonprofit.org/people/volunteers-and-your-organisation.

INE National Statistics Institute.

https://www.me.pt/xportal/xmam?xpgid=me main&xpid=INE&xlang=en

Lipovetsky, G.1988. The Age of Emptiness. Essays on Contemporary Individualism. Relógio de Água, Colecção Antropos, translation of the original 'L'Ere du Vide' (1985), by Miguel Serras Pereira and Ana Luisa Faria, Lisbon, 1st edition.

2009. Paradoxical Happiness. An Essay on the Society of Hyper-Consumption. Lisbon, Edições 70.

Lundqvist, J., de Fraiture, C. & Molden, D. 2008. Saving water: from field to fork - curbing losses and wastage in the food chain. SIWI Policy Brief, Stockholm International Water Institute (SIWI), Stockholm.

Ortiz, Renato.1994. Mundialisation and culture. 2.ed. São Paulo: Brasiliense

Parfitt, J., Barthel, M. & Macnaughton, S. 2010. Food waste within food supply chains: quantification and potential for change to 2050. Phil. Trans. R. Soc., vol. 365, pp. 3065-3081.

Putman, Robert D. (2002) Community and Democracy: the experience of modern Italy. 3 ed. Rio de Janeiro: FGV.

Re-Food 4 Good, 2017. Available at https://www.re-food.org/pt

Roberts, P. 2008. The End of Food. Boston: Houghton Mifflin.

Skoll World Forum. What is social entrepreneurship?, Available at: http://skoll.org/ Stuart T. 2009. Waste: uncovering the global food scandal. London: UK. Penguin Books. Schumpeter, J. 1985 "The Fundamental Phenomenon of Economic Development". In The Theory of Economic Development. Rio de Janeiro: Nova Cultural

ECA, European Court of Auditors, 2016. Fighting food waste: An opportunity for the EU to improve resource efficiency in the food supply chain. Luxembourg: European Court of Auditors. Accessed at https://www.eca.europa.eu/Lists/ECADocuments/SR16 34/SR FOOD WASTE EN .pdf

The Uniplanet, 2016.WeFood supermarket. Available at http://www.theuniplanet.com/2016/04/wefood-supermercado-na-dinamarca-que- so 26.html

Tukker, A., Huppes, G., Guinée, J., Heijungs, R., de Koning, A., Van Oers, L., et al. 2006. Environmental impact of products (EIPRO); analysis of the life cycle environmental impacts related to the final consumption of the EU-25. European Commission, DG JRC, Institute for Prospective Technological Studies, Technical report EUR 22284 EN. http://ec.europa.eu/environment/ipp/pdf/eipro report.pdf

UNEP, 2011. "Towards a Green Economy: Pathways to Sustainable Development and Poverty

Eradication". Available at www.unep.org/greeneconomy

UNRIC, 2015. Guide to sustainable development: 17 goals to transform our world. Brussels: United Nations Regional Information Centre for Western Europe. Accessed at https://www.unric.org/pt/images/stories/2016/ods 2edicao web pages.pdf

US EPA. "Food Recovery Hierarchy. Available at https://www.epa.gov/sustainable-management-food/food-recovery-hierarchy

MIX
Papier aus verantwortungsvollen Quellen
Paper from responsible sources
FSC® C105338

Printed by Books on Demand GmbH, Norderstedt / Germany